MARQUES VICKERS

CACTUS CADAVERS
Vanishing Phantoms Of Suburban Sprawl

CACTUS CADAVERS
Vanishing Phantoms of Suburban Sprawl

By Marques Vickers

MARQUIS PUBLISHING
BAINBRIDGE ISLAND, WASHINGTON

Copyright @2019-2024 Marques Vickers

All rights reserved. Copyright under Berne Copyright Convention, Universal Copyright Convention, and Pan-American Copyright Convention. No part of this book may be reproduced, stored in a retrieval system, or transmitted in any form, or by any means, electronic, mechanical, photocopying, recording, or otherwise, without prior permission of the author or publisher.

Version 1.2

Published by Marquis Publishing
Bainbridge Island, Washington

Vickers, Marques, 1957

CACTUS CADAVERS
Vanishing Phantoms of Suburban Sprawl

Dedication: To my daughters Charline and Caroline.

TABLE OF CONTENTS

Vanishing Phantoms of Suburban Sprawl
Vibrant Cactus
Wounded Cactus
Dead Cactus

About The Author

The Transformation of the American Desert

It has been speculated that the human history of the Salt River Valley, the terrain of contemporary Phoenix, commenced with nomadic paleo-Indians. These earliest civilizations inhabited the Americas during the final glacial episodes of the late Pleistocene period, approximately 6,000 BC.

Tribes hunted mammoths, mastodons, giant bisons, camels, and giant sloths that eventually migrated eastward. The initial nomadic tribes followed vacating the region. Tribes originating from Mexico to the south and California to the west would replace them.

Around 1,000 BC, a subsequent core of settlements would inhabit the territory. Corn farmers, builders, and permanent villagers would evolve into the Hohokam civilization. Within 500 years, the Hohokam culture had established an elaborate canal system enabling agriculture to flourish.

Around 1450, the Hohokam suddenly and mysteriously disappeared. By the 16th century arrival of Europeans, the O'odham and Sobaipuri tribes primarily inhabited the region.

The Spanish concentrated their settlements within southern Arizona and the Tucson area. American settlers first settled Central Arizona during the early 19th century. A military outpost to the east of current day Phoenix provided an administrative base for the community's agrarian base. Irrigation projects tamed the inhospitable desert and the local economy was based on cotton, citrus, cattle, and copper.

The availability of air conditioning to counter the oppressive summer dry heat stimulated a post-World War II population surge. The Phoenix metro area has increased in population an estimated average of 4% for the past forty years. Phoenix is the fifth largest city in American with projections that it may become the fourth within the next five years.

On the periphery of Scottsdale, Arizona is a troubling reminder of the consequences resulting from suburban expansion and desert encroachment. The silent victims are powerless to protest or alter the unblinking destiny of development. Gashed, wounded, and disfigured cacti litter the remaining vacant terrain, rapidly disappearing into subdivisions of residential housing tracts, strip malls, and commercial constructions.

This growth proliferates to accommodate a swelling and aging population migration seeking the warmer climate the Arizona desert can accommodate. One day, the cacti's diminishing and lost presence may be mourned once the transitional madness subsides. In the meanwhile, this edition illustrates the decline of these desert patriarchs.

It seems unimaginable that amidst the expansive desert landscape these icons could ever entirely vanish. Yet like the mammoths and Hohokam civilization from centuries past, adaptation for them becomes difficult if not impossible. Domestically cultivated cacti may only emulate the nobility of their freely born brethren that tower majestically amidst the desert landscape. Their declining numbers are emblematic of mankind's growing disharmony with his surrounding environment.

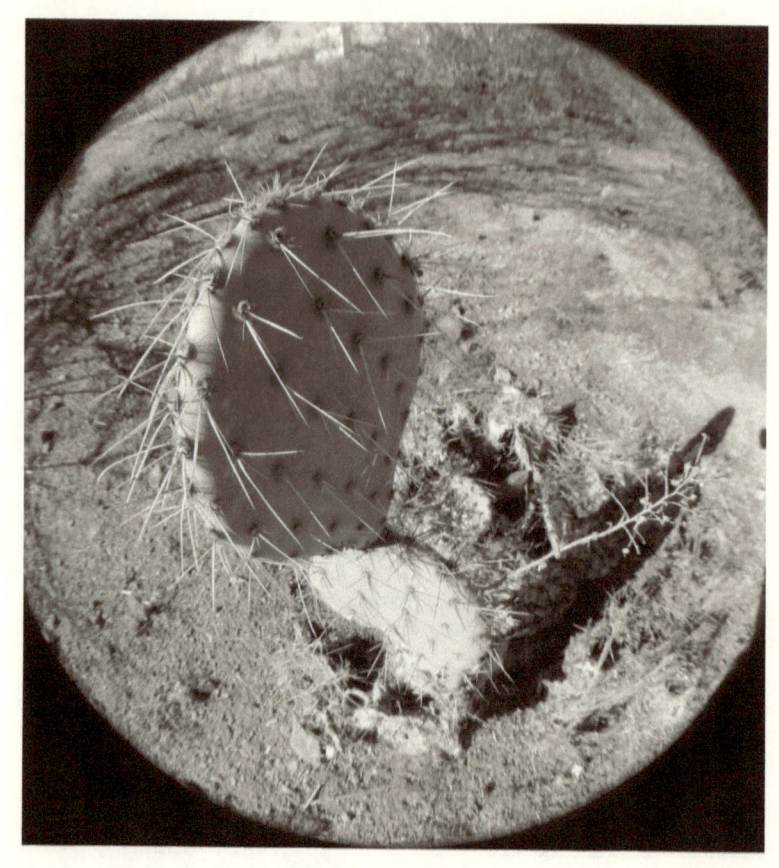

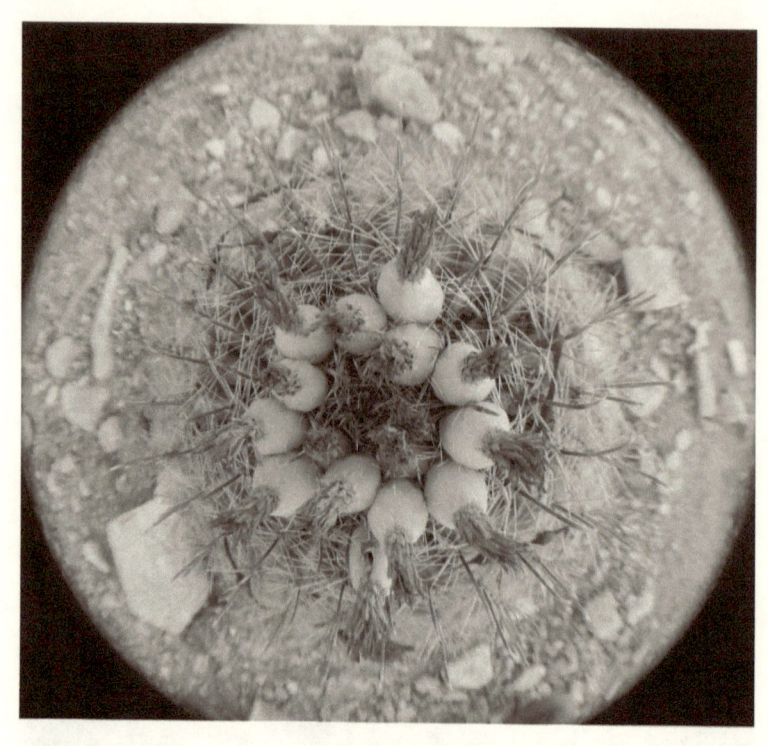

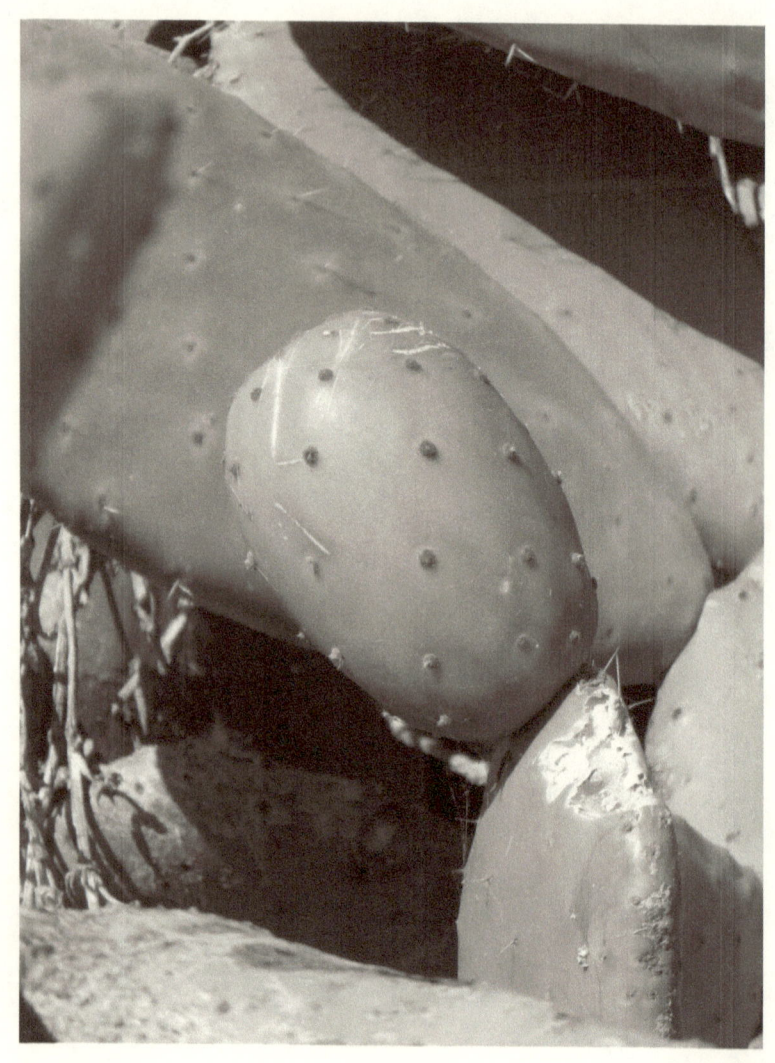

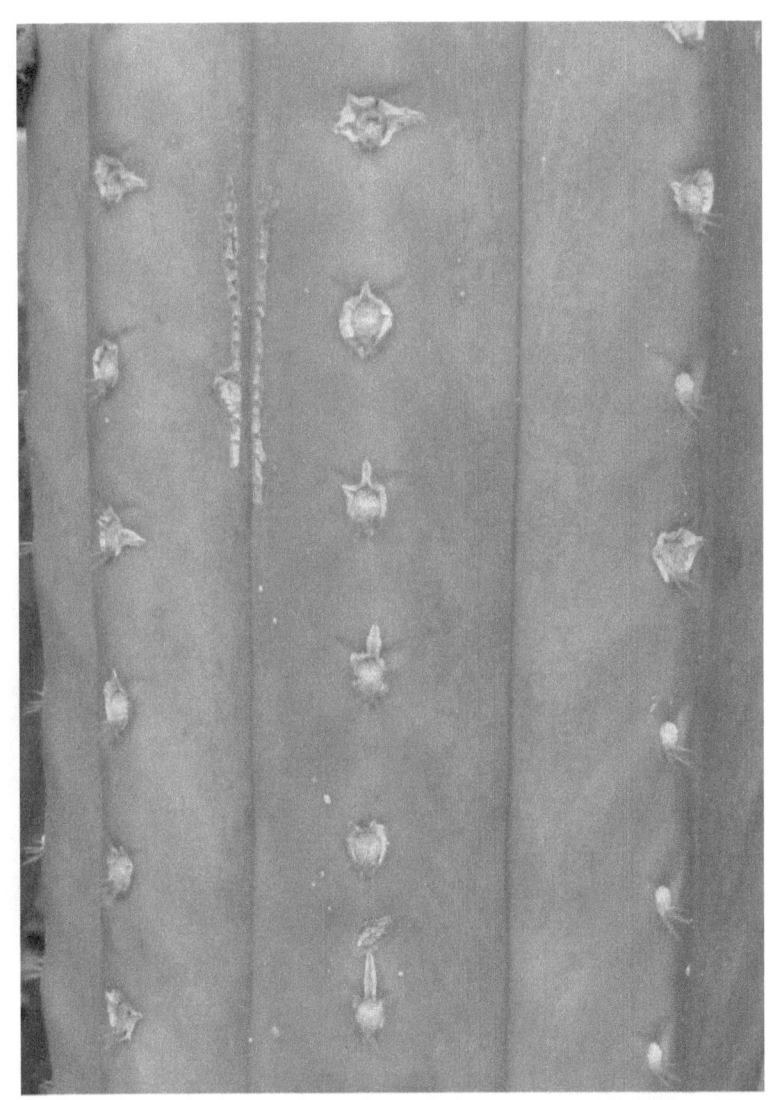

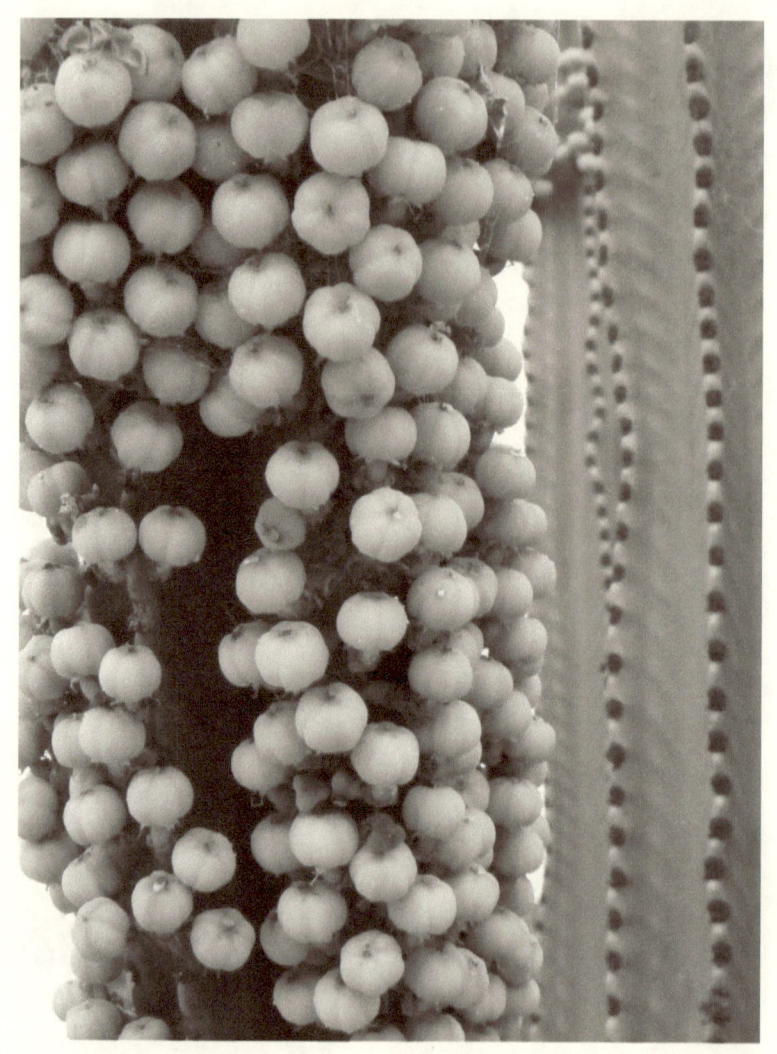

WOUNDED

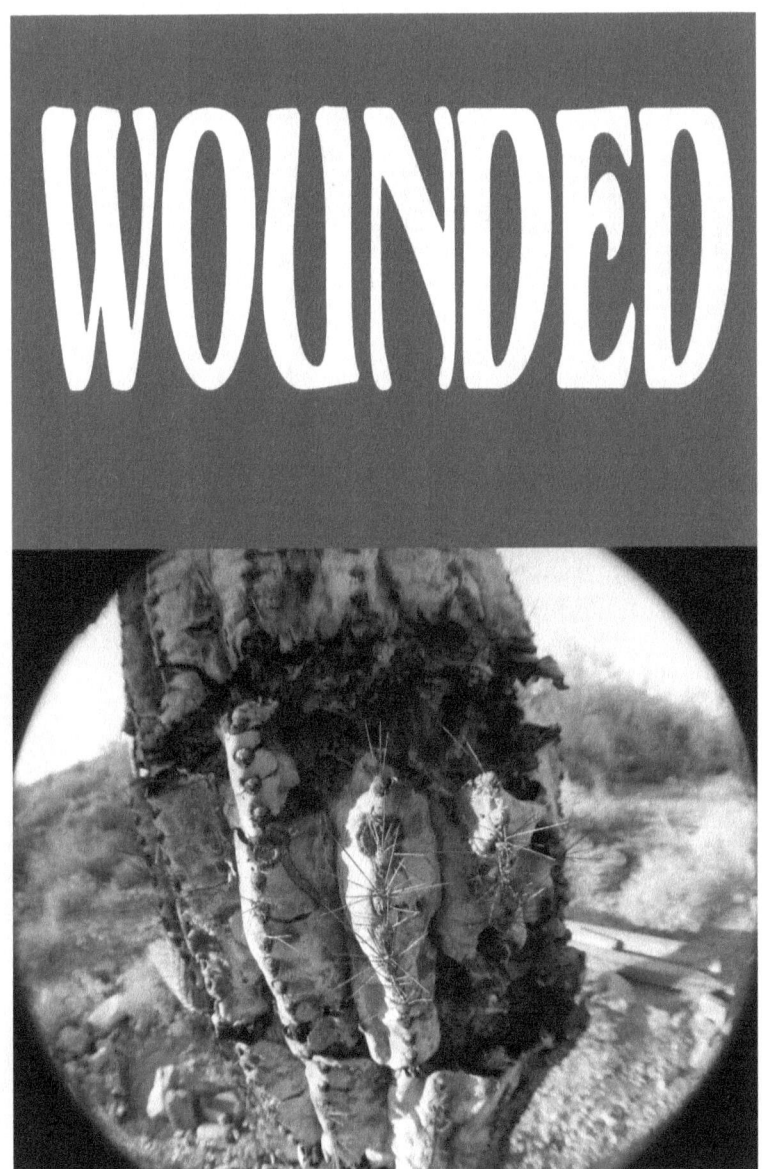

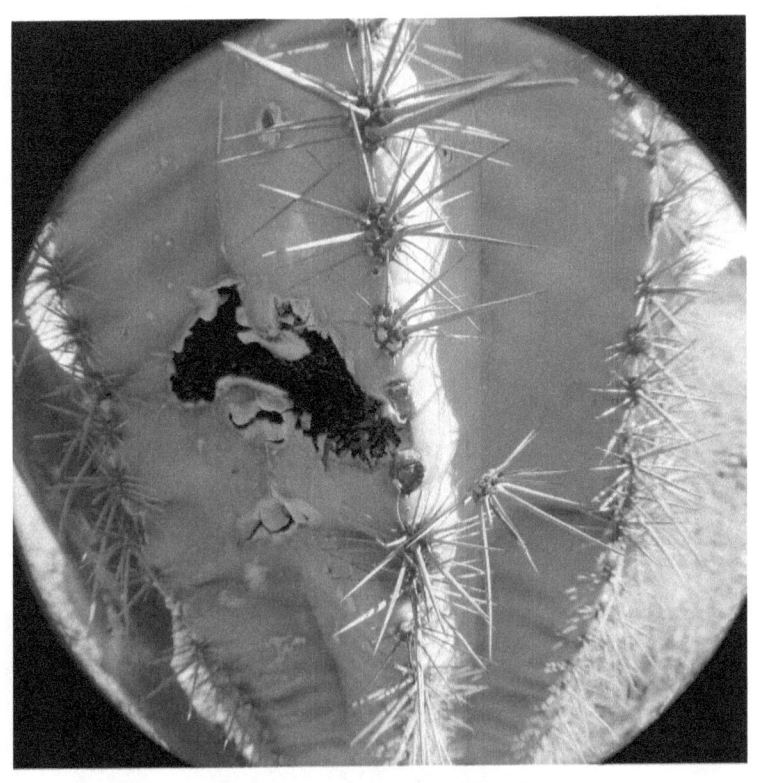

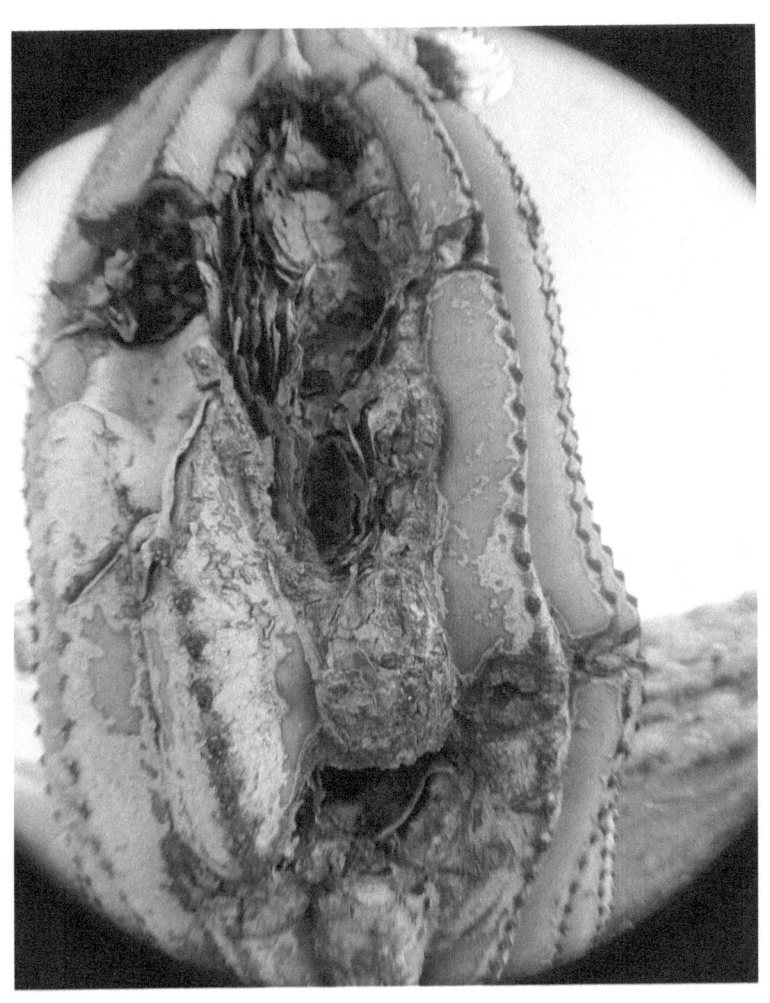

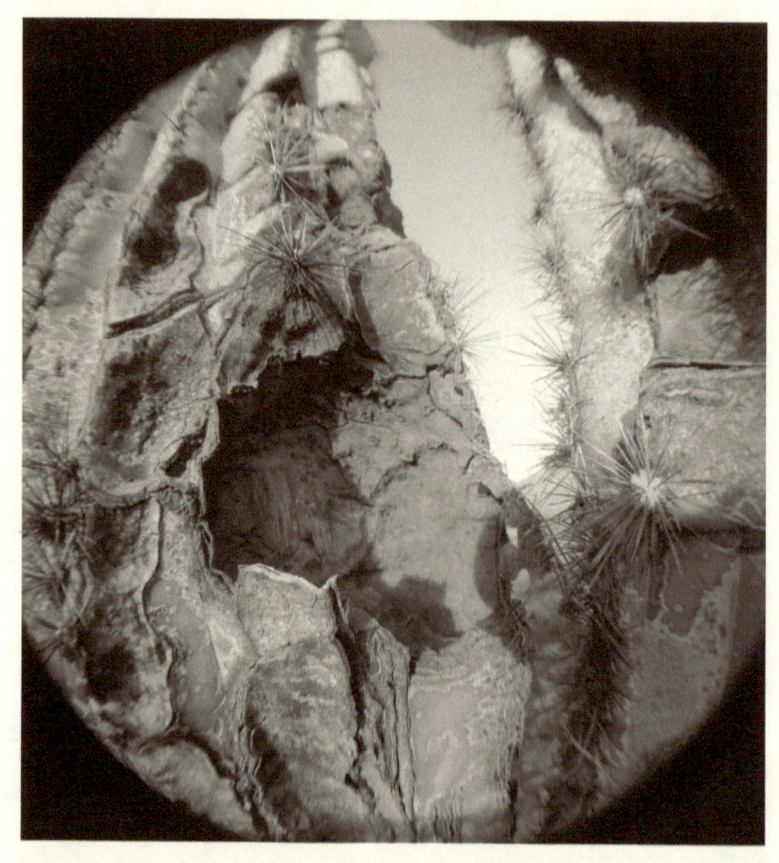

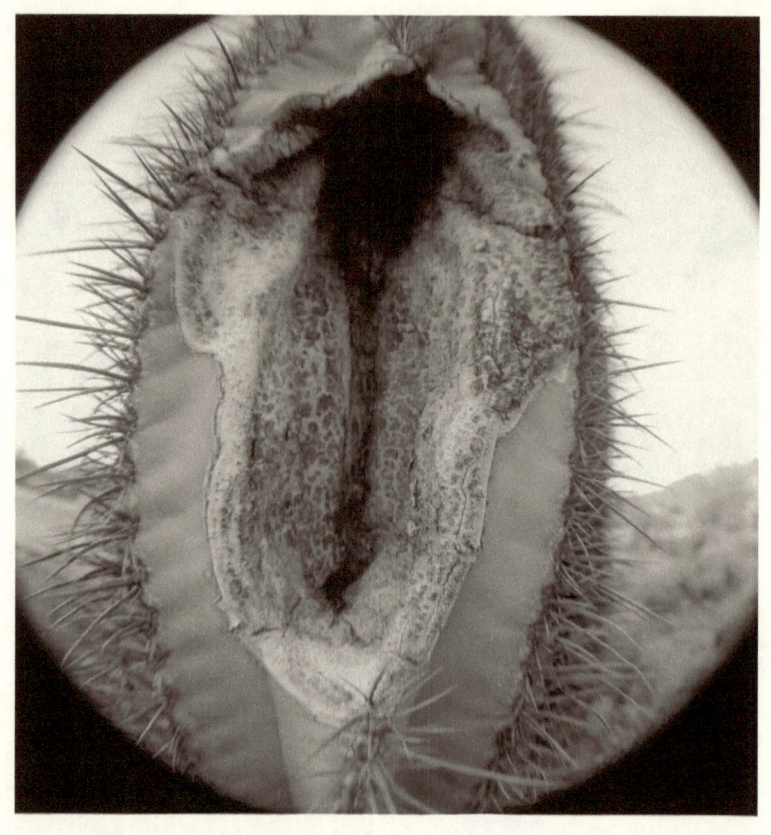

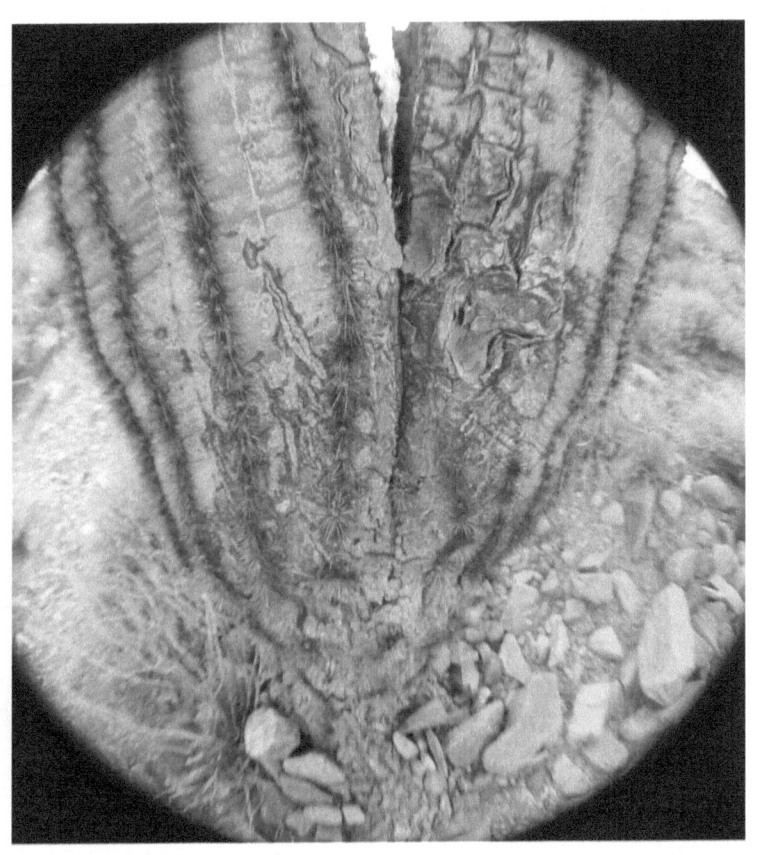

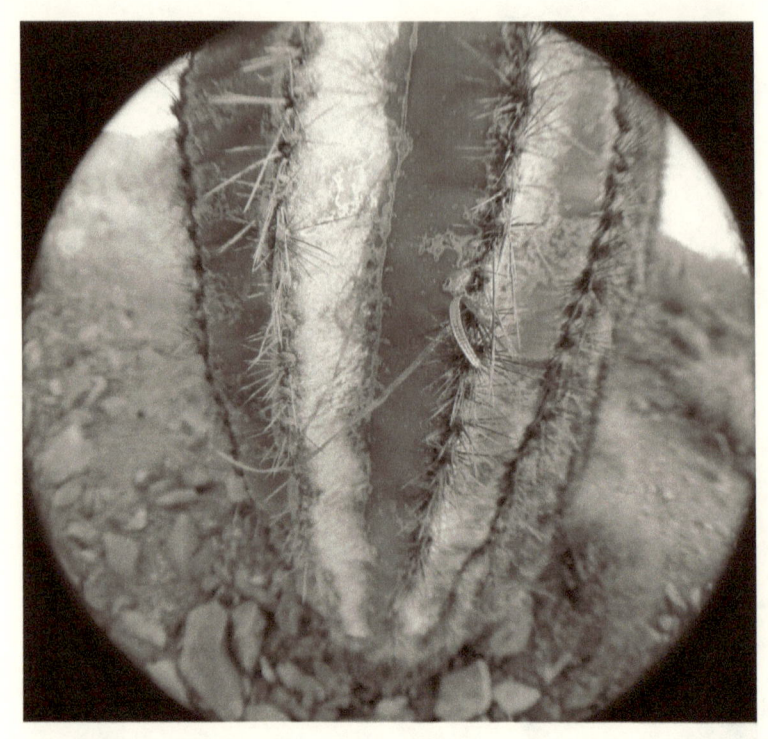

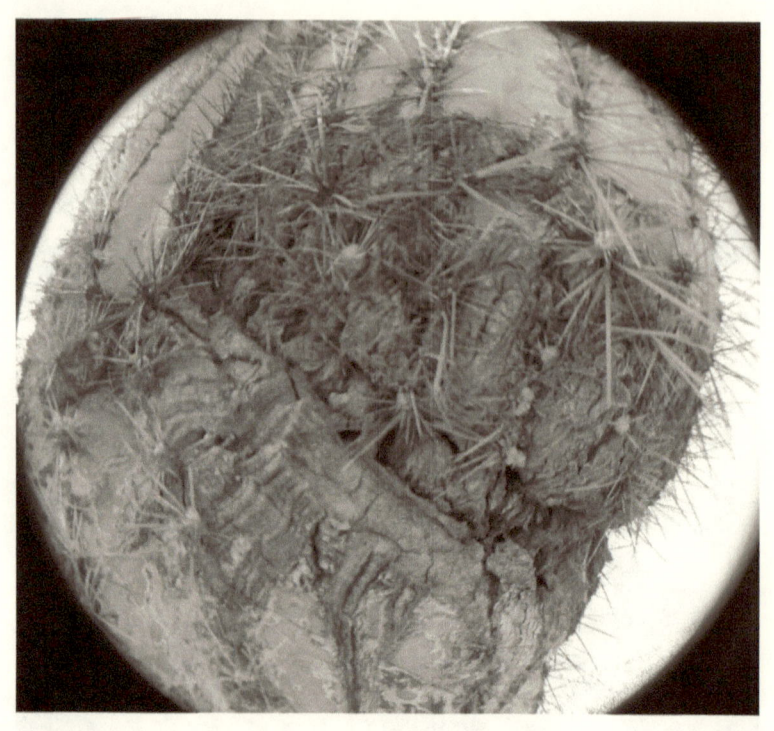

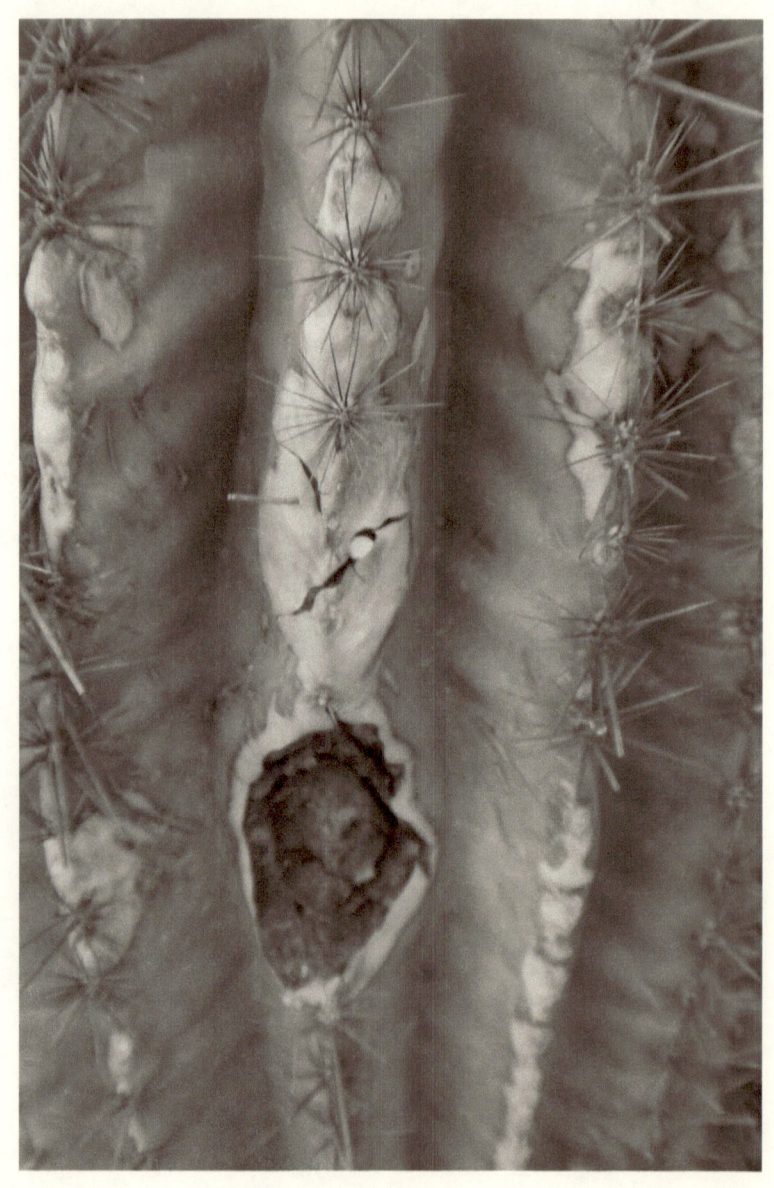

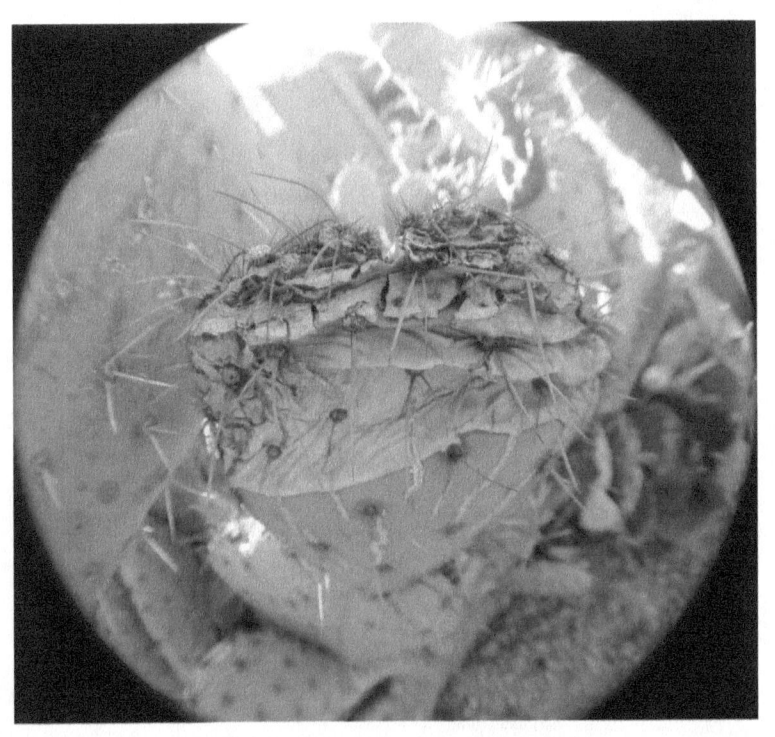

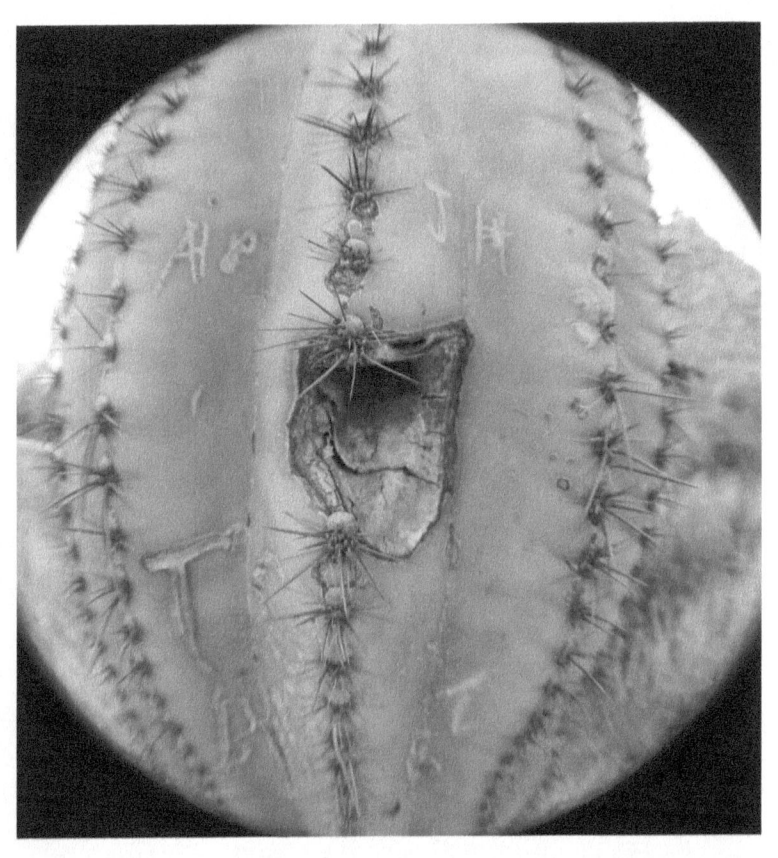

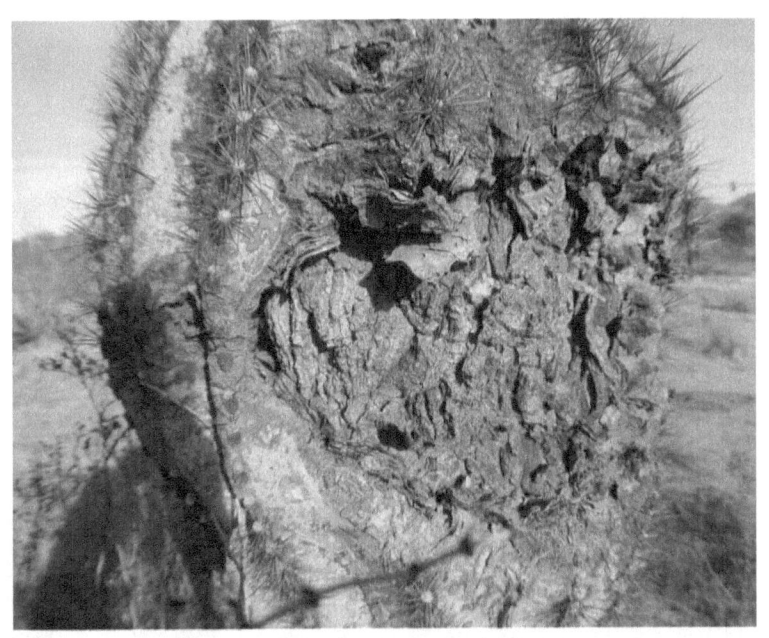

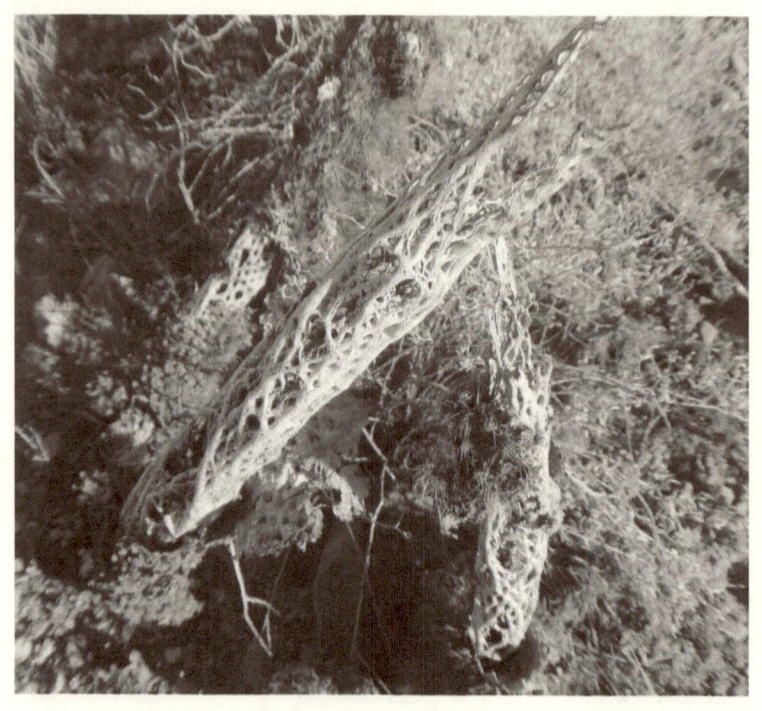

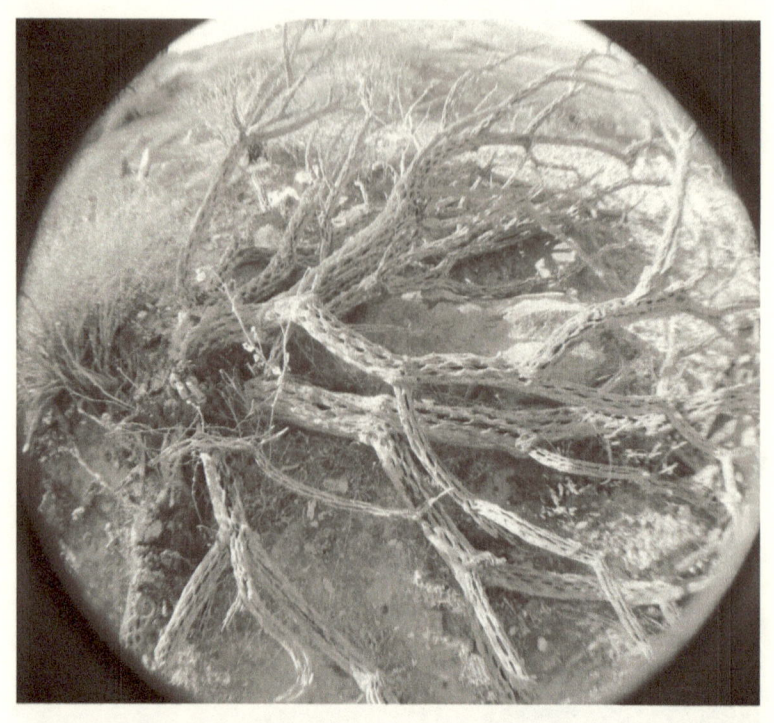

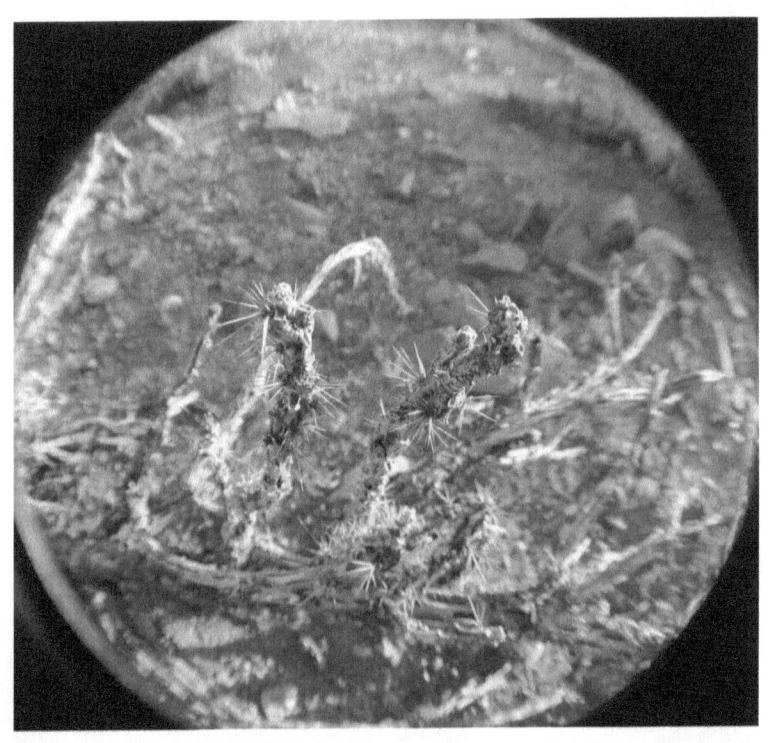

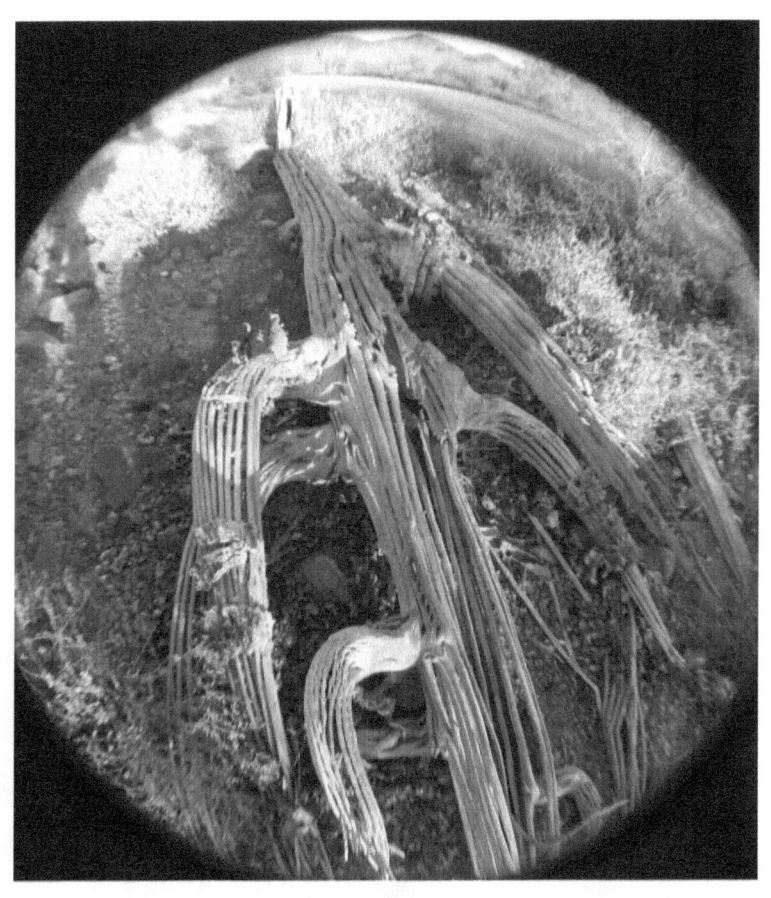

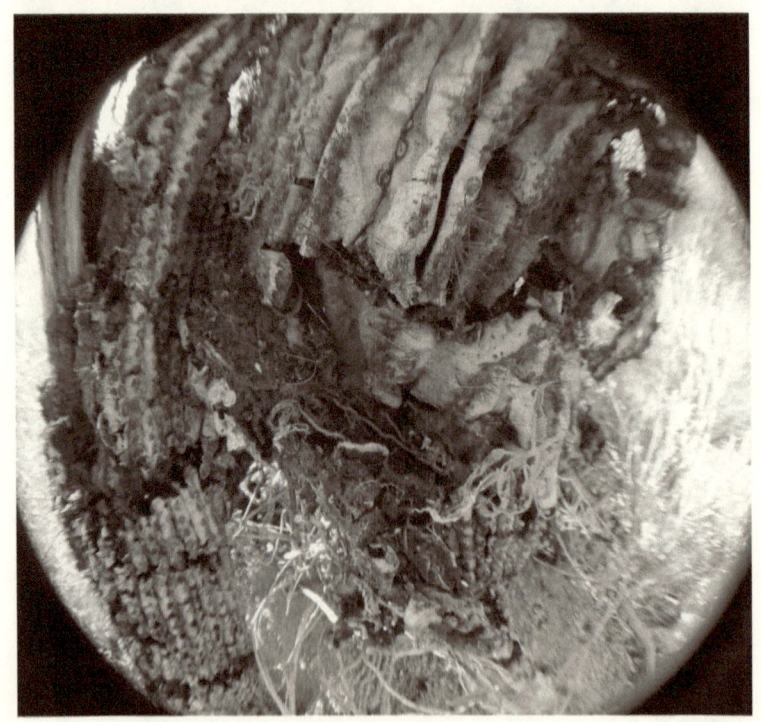

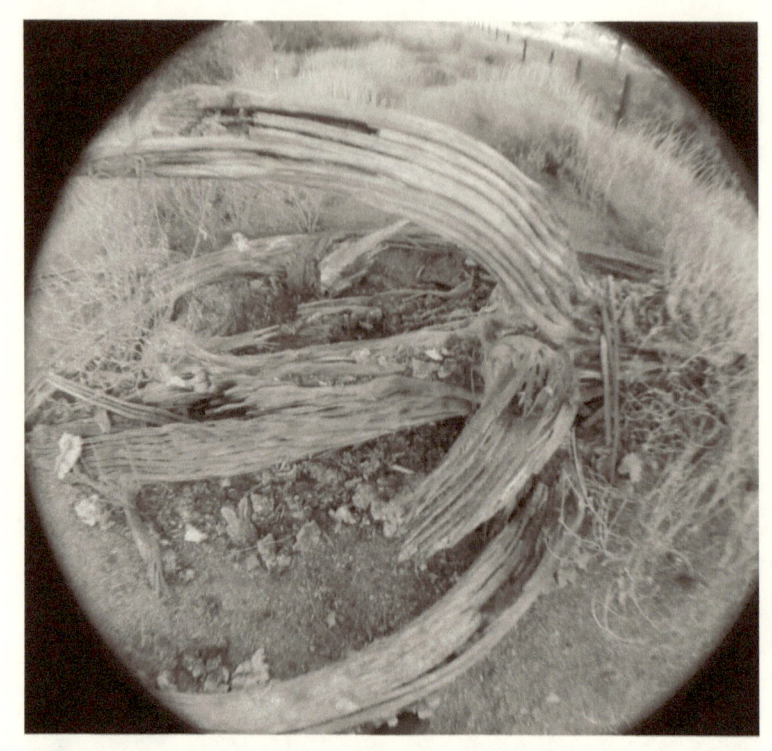

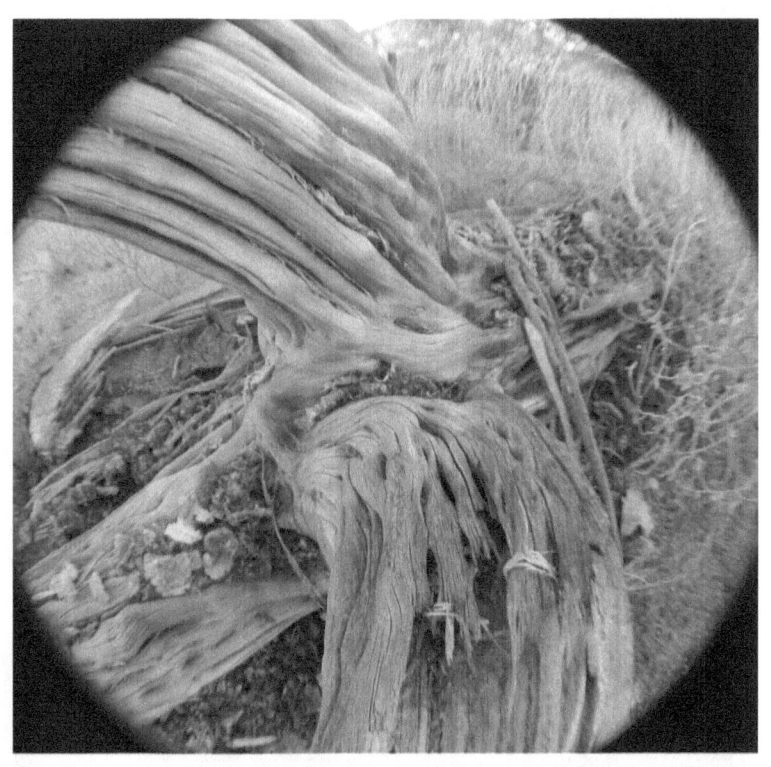

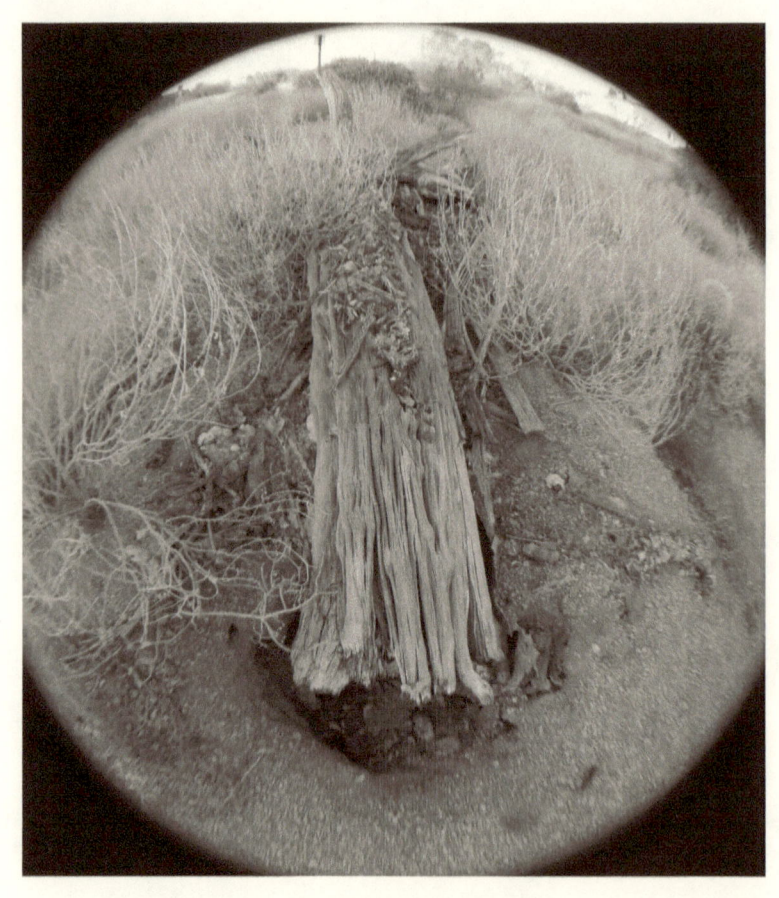

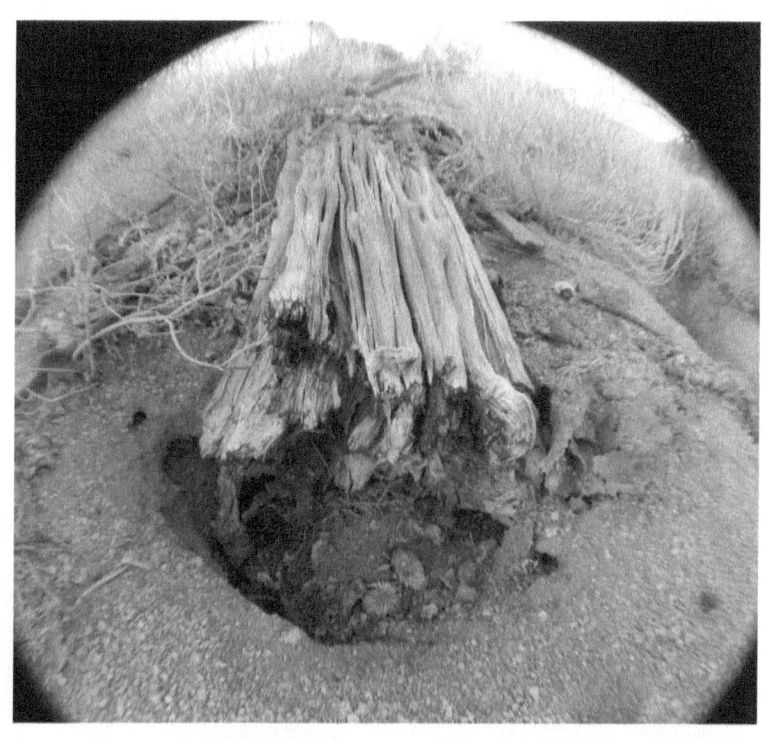

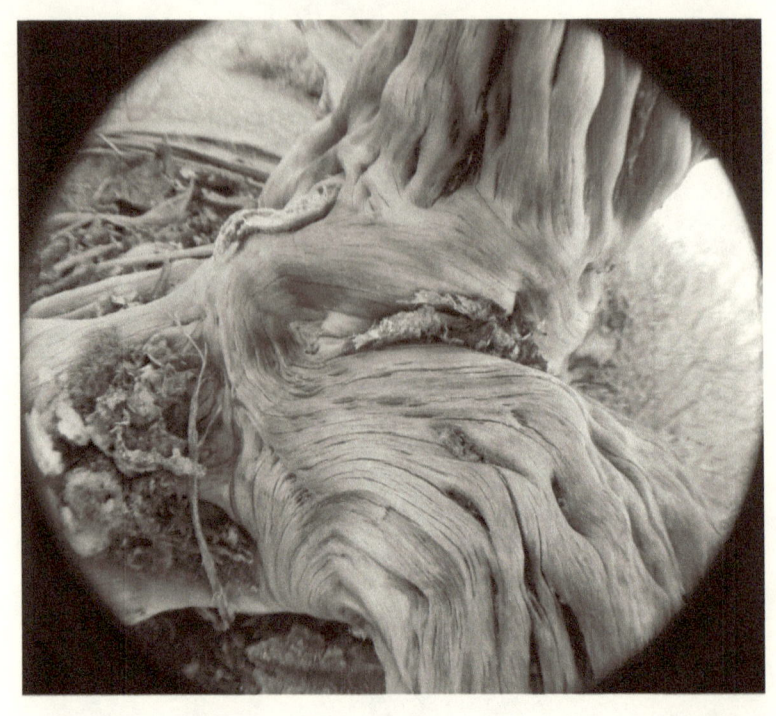

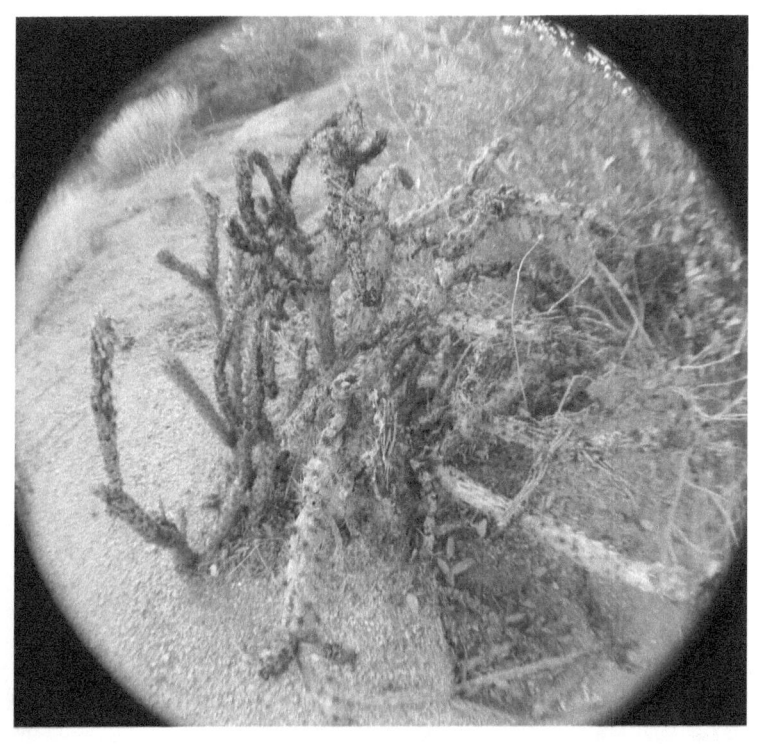

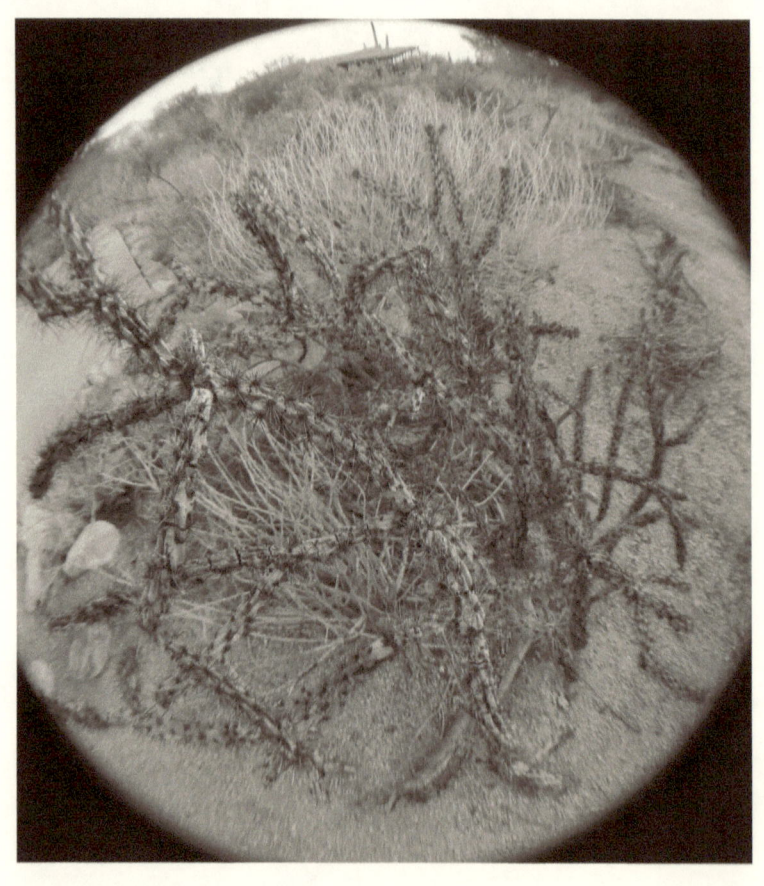

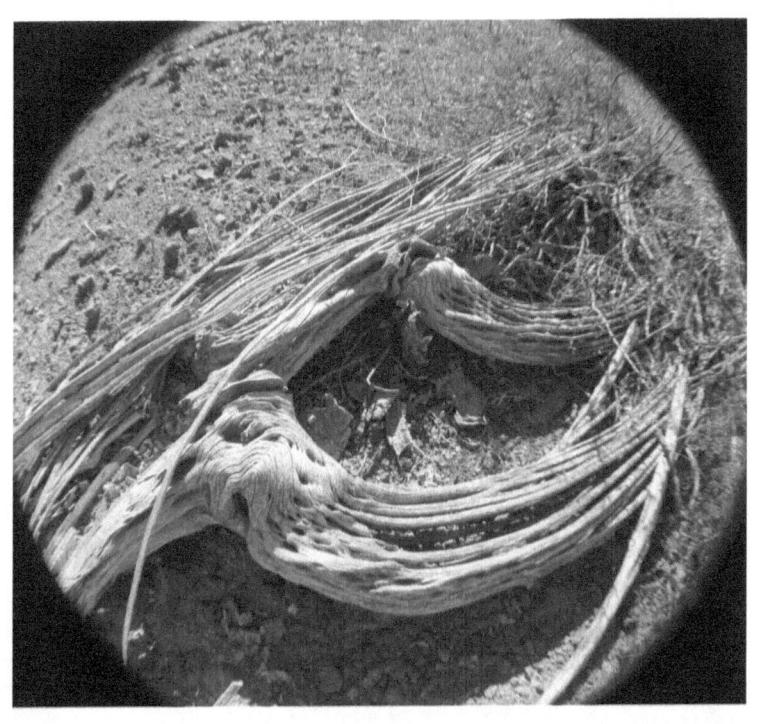

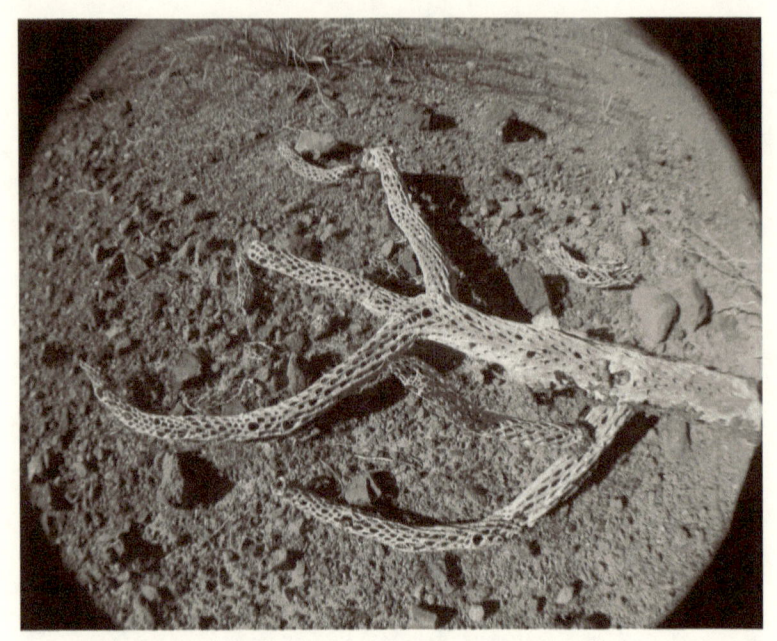

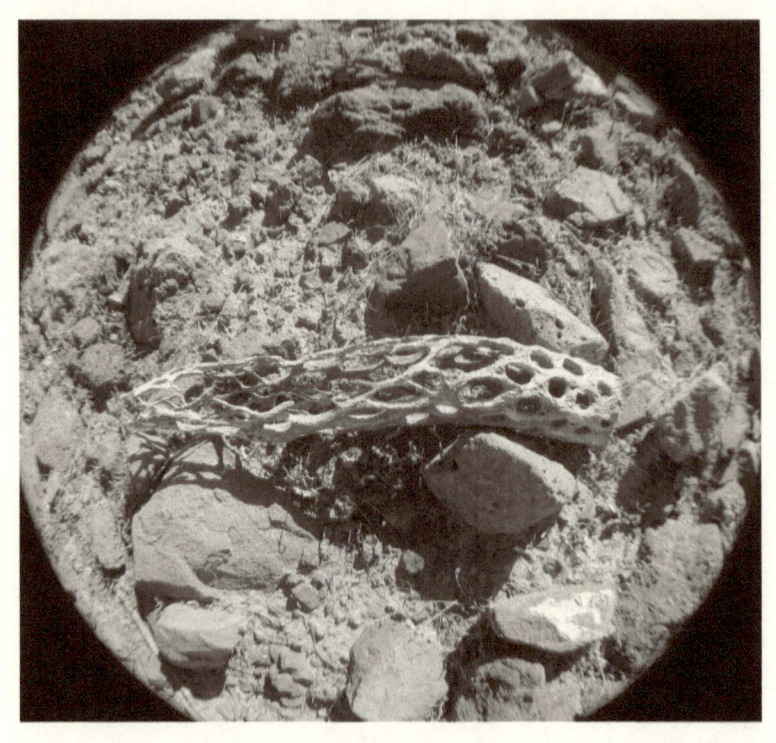

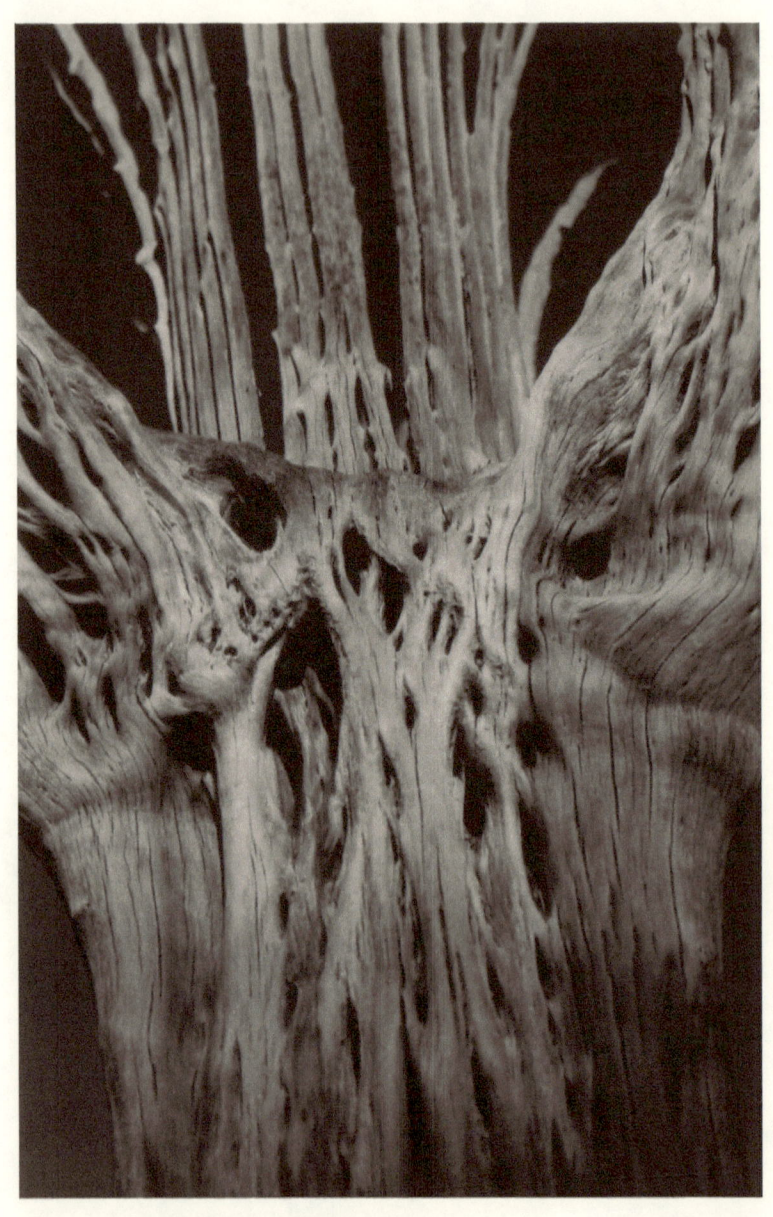

ABOUT THE AUTHOR

Author, photographer and visual artist Marques Vickers was born in 1957 in Vallejo, California. He graduated from Azusa Pacific University in Los Angeles and became the Public Relations and Executive Director for the Burbank, California Chamber of Commerce between 1979-84.

Professionally, he has operated travel, apparel, wine, rare book and publishing businesses. His paintings and sculptures have been exhibited in art galleries, private collections and museums in the United States and Europe. He has previously lived in the Burgundy and Languedoc regions of France and currently lives in the South Puget Sound region of Western Washington.

He has written and published over one hundred and forty books spanning a diverse variety of subjects including true crime, international travel, social satire, wine production, architecture, history, fiction, auctions, fine art, poetry and photojournalism.

He has two daughters, Charline and Caroline who reside in Europe.

www.ingramcontent.com/pod-product-compliance
Lightning Source LLC
Chambersburg PA
CBHW032004170526
45157CB00002B/532